the fib notebook
a fibonacci journal

AF797207

this notebook belongs to:

1.1.2.1.3.2.4.3.5.5.6.8.7.13.8.21.9.34.10.55.11.89.12.144.13.233.14.377.15.610.16.987.17.1597.18.2584.19.4181.20.6765.21.10946.22.17711.23.28657.24.46368

.25.75025.26.121393.27.196418.28.317811.29.514229.30.832040.31.1346269.32.2178309.33.3524578.34.5702887.35.9227465.36.14930352.37.24157817.38.3

9088169.39.63245986.40.102334155.41.165580141.42.267914296.43.433494437.44.701408733.45.1134903170.46.1836311903.47.2971215073.48.480752697

6.49.7778742049.50.12586269025.51.20365011074.52.32951280099.53.53316291173.54.86267571272.55.139583862445.56.225851433717.57.365435296162.

58.591286729879.59.956722026041.60.1548008755920.61.2504730781961.62.4052739537881.63.6557470319842.64.10610209857723.65.17167680177565.6

6.27777890035288.67.44945570212853.68.72723460248141.69.117669030460994.70.190392490709135.71.308061521170129.72.498454011879264.73.80651

5533049393.74.1304969544928657.75.2111485077978050.76.3416454622906707.77.5527939700884757.78.8944394323791464.79.14472334024676221.80.2

3416728348467685.81.37889062373143906.82.61305790721611591.83.99194853094755497.84.160500643816367088.85.259695496911122585.86.42019614

0727489673.87.679891637638612258.88.1100087778366101931.89.1779979416004714189.90.2880067194370816120.91.4660046610375530309.92.7540113

804746346429.93.12200160415121876738.94.19740274219868223167.95.31940434634990099905.96.51680708854858323072.97.83621143489848422977.98

.135301852344706746049.99.218922995834555169026.100.354224848179261915075.101.573147844013817084101.102.927372692193078999176.103.1500

520536206896083277.104.2427893228399975082453.105.3928413764606871165730.106.6356306993006846248183.107.10284720757613717413913.108.16

641027750620563662096 109.26925748508234281076009.110.43566776258854844738105.111.70492524767089125814114.112.114059301025943970552219

.113.184551825793033096366333.114.298611126818977066918552.115.483162952612010163284885.116.781774079430987230203437.117.12649370320429

97393488322.118.2046711111473984623691759.119.3311648143516982017180081.120.5358359254990966640871840.121.8670007398507948658051921.12

2.14028366653498915298923761.123.22698374052006863956975682.124.36726740705505779255899443.125.59425114757512643212875125.126.96151855

463018422468774568.127.155576970220531065681649693.128.251728825683549488150424261.129.407305795904080553832073954.130.65903462158763

0041982498215.131.1066340417491710595814572169.132.1725375039079340637797070384.133.2791715456571051233611642553.134.4517090495650391

871408712937.135.7308805952221443105020355490.136.11825896447871834976429068427.137.19134702400093278081449423917.138.309605988479651

13057878492344.139.50095301248058391139327916261.140.81055900096023504197206408605.141.131151201344081895336534324866.142.21220710144

0105399533740733471.143.343358302784187294870275058337.144.555565404224292694404015791808.145.898923707008479989274290850145.146.1454

489111232772683678306641953.147.2353412818241252672952597492098.148.3807901929474025356630904134051.149.6161314747715278029583501626

149.150.9969216677189303386214405760200.151.16130531424904581415797907386349.152.26099748102093884802012313146549.153.422302795269984

66217810220532898.154.6833002762909235101982253367944 7.155.11056030715609081723763275421 2345.156.1788903347851831682574552878917 92.15

7.28945064194127398549508804210413 7.158.46834097672645715375254332999592 9.159.7577916186677311392476313721000 66.160.12261325953941882

93000174702095995.161.19839242140619194322478060741960 61.162.32100568094561077252479807762920 56.163.51939810235180271574957868504881

17.164.84040378329741348827437676267801 73.165.13598018856492162040239554477268290.166.22002056689466296922983322104048463.167.3560007

5545958458963222876581316753.168.57602132235424755886206198685365216.169.932022077813832148494290752666819 69.170.1508043400168079707

35635273952047185.171.24400654779819118558506434921872915 4.172.39481088781499915632069962317077633 9.173.638817435613190341905763972 3

89505493.174.103362832342818949822646359556028183 2.175.167244575904137984013222756794978732 5.176.27060740824695693383586911635100691

57.177.4378519841510949178490918731459856482.178.70845939239805185168496098949699256 39.179.11463113765491467695340528626429782121.180

.18547707689471986212190138521399707760.181.30010821454963453907530667147829489881.182.485585291444354401197208056692291976 41.183.785

6935059939889402725147281705868752 2.184.12712787974383433414697227848628788516 3.185.20569723034323322817422375130334657268 5.186.3328

25110087067562321196029789634457848.187.53852234043030079049541978109298103053 3.188.87134745051736835281661581088261548838 1.189.1409

8697909476691433120355919755965189 14.190.22812172414650374961286514028582120072 95.191.36910870324127066394406869948338085262 09.192.5

972304273877744135569338397692020533504.193.9663391306290450775010025392525829059713.194.15635695580168194910579363790217849593217.

195.25299086886458645685589389182743678652930.196.4093478246662684059616875297296152824614 7.197.66233869353085486281758142155705206

899077.198.10716865181971232687792689512866673514 5224.199.17340252117279781315968503728437194204 4301.200.28057117299251014 0037611932

413038677189525.201.45397369416530795319729696969741061923382 6.202.73454486715781809323490890211044929642335 1.203.118851856132312604

643220587180785991565717 7.204.19230634284809441396671147739183092120805 28.205.311158198980407018609932064572616912773770 5.206.503464

54182850143257664354196444783398182 33.207.81462274080890845118657560653706474675559 38.208.13180872826374098837632191485015125807374

171.209.21327100234463183349497947550385773274930109.210.34507973060837282187130139035400899082304280.211.5583507329530046553662808 6

585786672357234389.212.90343046356137747723758225621187571439538669.213.14617811965143821326038631220697424379677305 8.214.2365211660

0757596098414453782816181523631172 7.215.38269928565901417424453085003513605903308478 5.216.61922045166659013522867538786329787426939

6512.217.100191973732560430947320623789843393330248129 7.218.16211401889921944447018816257617318075718778 09.219.262305992631779875417

508786366016574087435910 6.220.4244200115309993198876969489421897548446236915.221.6867260041627791953052057353082063289320596021.222

111114601569377851519290268425039608377668329 36.223.17978720198565577104981084195586024127087428957.224.290901803555033622569101110 3

8089984964854261893.225.47068900554068939361891195233676009091941690850.226.76159080909572301618801306271765994056795952743.227.1232

2798146364124098069250150544200314873764359З.228.199387062373213542599493807777207997205533596336.229.32261504383685478358018630928

2650000354271239929.230.522002106210068326179680117059857997559804836265.231.844617150046923109759866426342507997914076076194.232.13

6661925625699143593954654340236599547388091 2459.233.22112364063039145456994129697448739933879569886 53.234.35778556625609059816389 59

5131472399888618379011 12.235.57890920688648205273383724828921139822497948897 65.236.93669477314257265089773319960393539711116327908 77

.237.15156039800290547036315704478931467953361427680642.238.24522987531716273545293036474970821924473060471519.239.39679027332006820

58160874095390228987783448815 2161.240.64202014863723094126901777428873111802307548623680.241.103881042195729914708510518382775401 68

0142036775841.242.16808305705945300883541229581164851348244958539952 1.243.27196409925518292354392281419442391516259162217536 2.244.44

00471563146359323793351100060724286450412075748 83.245.712011255569818855923257924200496343807632829750245.246.11520584118844547883 0

2593034206568772452674037325128.247.186406966745427364422585095840706511626030686707537 3.248.30161280793387284325284439926136338887

12980904400501.249.48801977467930020767542949510206990049732877714758 74.250.7896325826131730509282738943634332893686268675876375.25

1.127765235729247325860370338946550318986595564473522 49.252.2067284939905646309531977283828936479234582512322862 4.253.33449372971981

19568135680673294439669100538157058087 3.254.5412222237103765877667657957123376148335120669380949 7.255.8757159534301885445803338630 4

178158174356588264390370.256.14169381771405651323470996587541191965770779495819986 7.257.22926541305707536769274335217959007783206 43

83222590237.258.370959230771131880927453318055001997489772178180790104.259.600224643828207248620196670234592075321836561403380341.26

0.97118387459933912954764998828959407281160873958417044 5.261.1571408518427546378167846658524186148133445300987550786.262.2542592393

26885507715496646813780220945054040571721231.263.41140009114544318858833433053379663690784993415592720 17.264.6656593304481317393598

839952151746590023553382130993248.265.10770594215935749279482183257489712959102052723690265265.266.1742718752041706667308102320964

4595491256061058212585 13.267.281977817363528159525632064671311725082276588295115237 78.268.45624969256769882625644229676772632057353

264935332782291.269.73822750993122698578207436143903804565580923764844306069.270.119447720249892581203851665820676436622934188700 17

7088360.271.193270471243015279782059101964580241188515112465021394429.272.31271819149290786098591076778525667781144930116519848278

273.5059886627359231407679698697498369189999644136302198772 18.274.8187068542288310017538806375350935968114137147954183600 07.275.1324

69551696475414252185050728493051581137812842563823 7225.276.21434023711935851442757311448200241126227918432210565972 32.277.346809788

815833928679758165210495462843416997164669483445 7.278.56115002593519244310733127969249787410569618148677514316 89.279.907959814751C2

63717870894449029933369491131786514446266146.280.14691098406862188148944207245954912110548093601382197697835.281.237706965543724518

66815101694984845480039225387896643963981.282.384617949612346400157593089409397575905873189892788416618l6.283.622324915156070918825

74410635924603070626544377175485625797.284.1006942864768417318983337195768643606612138633664543272876l3.285.16292677799244882378090

8130212788963731840407743629812913410.286.263621064469290555679241849789653324393054271110084140201023.287.426547842461739379460149

98000244228812489467885371395311443 3.288.69016890693102993513939182979209561251794894996379809331545 6.289.11167167493927693l4599541

80979453790064284362881751204642988 9.290.180688565632379924973893363958663351316079257878131013974534 5.291.29236024057165685643384 7

544938117141380363620759882218617523 4.292.47304880620403678140774090889678049269644287863801323259205 79.293.765409046775693637841 58

845383489763407680649939789545120958l3.294.12384578529797304192493293627316781267732493780359086838016392.295.200386689975542405709

091781656657576085005587743380413501122 05.296.32423247527351544763402471792982538876233052554697128188128597.297.524619165249057853

34311649958648296484733611329035169538240802.298.848851640522573300977141217516308353609666638837322977263693 99.299.137347080577163

11543202577171027913184570027521276746726461020 1.300.22223224462942044552973989346190996720666939096499764990979600.301.3595793252

0658356096176566517218909905236721430926723225558980 1.302.58181156983600400649150555863409906625903415340576699724656940 1.303.94139

089504258756745327122380628816531140136771503422950215920 2.304.15232024648785915739447767824403872315704355211208012267487286 03.305

.246459335992117914139804800624667539688183688883583545625088780 5.306.3987795824799770715342824788687062628452272409956636682999616

408.307.6452389184720949856740872794933738025334109298792472139250504213.308.1044018500952072057208369758362080065378638170874910 88

22250120621.309.168925741942416704288245703785545386791204910075415809615006248 34.310.2733275920376239100090826796217533933290 68727

16290689783750745455.311.44225333398004061429732838340729878012027363723832270745251370289.312.715580926017664524306411063029052173

449342364401229605290021157 44.313.115783425999770513860373944643635095356961600163955231274253486033.314.18734151860153696629101505

094654031270189583660407819180325560177 7.315.30312494460130748015138899559017540805885743676803342307750908781 0.316.490466463202844

4464424040465367157207607532733721116148807646895 87.317.793591407804151926593793042126891128819610710140145037958273777397.318.1284

057871006996373036197088663606849580363983512256652839038466984.319.207764927881114829962999013079049797839997469365240169079731 22

44381.320.336170714981814467266618721945410482798033867716465834363635071136 5.321.54393564286292929722961773502446028063803133708 17

060034433662955746.322.88010635784474376449623645696987076343606520479817183780700136671 11.323.142404200070767306172585419199433104

407409654187987784125036766228 57.324.2304148358552416826222090648964201807510161746678049679057369028996 8.325.3728190359260089887 94

7944840958532851584258288557927520307736691282 5.326.6032338717812506714170035489922734659094420035235977199365105720 2793.327.976052

907707259660211798033088126751067867832379390471967284241156 18.328.1579286779488510331628801582080400216977309835902988191903794 81

318411.329.2555339687195769991840599615168526968045177668282378663871079054340 29.330.4134626466684280323469401197248927185022487504

1853668557748738675244 0.331.6689966153880050315310000812417454153067665172467745519645952921864 69.332.1082459262056433063877940200 9

666381338090152676653112375420826789389 09.333.1751455877444438095408940282208383549115781784912085789506677971125378.334.2833915139

50087115928688048317502168292479705257739702704876065006428 7.335.4585371016945309254695820765383405232040578837489482816555438621 1

89665.336.7419286156446180413982701248558426914965375890066879843604199271253952.337.1200465717339148966867852201394183214700595472

755636266015963789244361 7.338.1942394332983767008266122326250025906197133061762324250376383716369756 9.339.3142860050322915975133974

527644209120897728534517960516392347505614118 6.340.508525438330668298340009685389423502709486159628028476676873122198387 55.341.8228

1144336295989585340713815384441479925901307982452831610787275979941.342.1331336881693628194193416823543267917508745172707853004992

98099495818696.343.215414832505658809004682396169711233230800418578767753330908886771798637.344.3485485206750216284240240785240380 2

498167493584955305383020698626761733 3.345.563963353180680437428706474693749258212475354428320807161115873039415970.346.912511873855

7020658527305532177872831941502902778738609913228593070333303.347.14764752270363825032814370279115365414066256447061946681524387323

46449273.348.238898710089208456913416758112932382460077593498406852914376159165348257.6.349.3865462327928467072415604609040860366007

401579690263197296200323999931849.350.6254449428820551641549772190170184190608177514674331726439961915653414425.351.101199117567490

18713965376799211044556615579094364594923736162239653346274.352.163743611855695703555151489893812287472237566090389266501761241553

06760699.353.26494272942318589069480525788592273303839335703403521573912286394960106973.354.428686341278881594249956747779735020510

63092312442448224088410550266867672.355.69362907070206748494476200566565775354902428015845969798000696945226974645.356.112231541198

094907919471875344539277405965520328288418022089107495493842317.357.18159444826830165641394807591110505276086794834413438782008960

4440720816962.358.293825989466396564333419951255644330166833468672422805842178911936214659279.359.475420437734698220747368027166749

382927701417016557193662268716376935476241.360.769246427201094785080787978422393713094534885688979999504447628313150135520.361.1244

666864935793005828156005589143096022236302705537193166716344690085611761.362.20139132921368877909089439840115368091167711883945171

92671163973003235747281.363.3258580157072680796737099989600679905139007491100054385837880317693321359042.364.5272493449209568587646

043973612216714255778679494571578509044290696557106323.365.85310736062822493843831439632128966193947861705946259643469246083898784

65365.366.13803567055491817972029187936825113333650564850089197542855968899086435571688.367.2233464066177406735641233190003800995304535102068382350720289350747631403705 3.368.36138207717265885328441519836863123286695915870773021050058862406562749608741.369.584728

48379039952684853851736901133239741266891456844557261755914039063645794.370.946110560963058380132953715737642565264371827622298656

07320618320601813254535.371.15308390447534579069814922331066538976617844965368671016458237423464087690032 9.372.24769496057165162871

14445948844296462926156324159165757719029925552426901548 64.373.40077886504699741940959381819509503605879408206960328593648536678 98

83567055193.374.64847382561864904812103841307952468235140971448551986170838835934512625721005 7.375.104925269066564646753063223127 46

19718410203796555123147644873726135009824265250.376.1697726516284295515651670844354144400761613511040643009353262085480136081475 30

7.377.274697920694994198318230287562876411917181730759576615699813581161514590574055 7.378.444470572323423749883397351998290851993 34

3081863640916635139789709528198721586 4.379.71916849301841794820162763956116726391052481262321753233495337087104278929564 21.380.1163

639065341841698085024991559458115903867894486858448970093160580570988017228 5.381.188280755836025964628665263112062537981439270711 0

07598130504653145161377731287 06.382.304644662370210134437167762268008349571826060159693443027513969203218476533009 91.383.4929254182

06236099065833025380070887553265330870701041158018622348379854264296 97.384.797570080576446233503000787648079237125091391030394484 1

85532591551598330797306 88.385.1290495498782682332568833813028150124678356721901095525343551213899978185061603 85.386.208806557935912

85660718346006762293618034481129314900095290838054515765158589 1073.387.33785610781418108986406684137043794864818048348325855348726

350193515547009 2051458.388.54666266575009394647125030143806088482852529477640755444017188248031312 1677942531.389.884518773564275036

335317142808498833476705778259666107927435384415468591769993989.390.143118143931436898280656744424655971830523107303607366236760 72

668957817134479 36520.391.2315700212878644019141884587055058551781936851295739770295042651311250305217930509.392.374688165219301300

94845203130161827008716792433181343266264991820703201866586702 9.393.6062581865071657021090336618356676821869104775627553202957692 5

695182823238837975 38.394.980946351726467002303878864965829509195627269995936663562034248772531434254966456 7.395.158720453823363270 4

4129125268014971913825377475586919838578035057243596666433462105.396.2568150889960099706716791391767326700578165017554628647419837

754496891100896 3126672.397.41553554281937324111297039185688238919607027651133206312776412602212507675416588777.398.672350631815383 21

1784649531033615059253886778266794927869747901471814186843997154 49.399.1087886174634756452897619922890497448449957054778126990997 5

1202749393926359816304226.400.1760236806450139664682269453924112507703843833044921918867259928965753450442160196 75.401.28481229810 8

489611757988937 6814609956153800887823048909864771956459692714040323239 01.402.46083597875350357822621588307387224638576447208679708

287320318854254461644824834357 6.403.745648276861993189984204820755333242001144560869101973859680384188513887852280667477.404.120648

42556154967682104207038292054883869090329558990567328835727310585043005290110 53.405.19521325324774899581946255245845387303880535 93

825001030592563956919572392152809678530.406.315861678809298672640504622841374421877496262678090008732544752965063089645333868958 3.

407.511074932057047668459967175299828294916301622060590111791801148657020328860614836811 3.408.826936610866346341100471798141202716 7

937978847386801205243459016220834185059487057696.409.133801154292339400956043897344103101171009950679927023231614705027910374736 65

635425809.410.216494815378974035066091077158223372850389739153795035284049295190118716587251224835 05.411.350295969671313436022134 97

450232647402139968983372205851566400021802909132390757909314.412.566790785050287471088226051660549846871789428987517093799713295 40

81478079111588039281 9.413.917086754721600907110361026162876320893189118821239152315377295626176899235066383021 33.414.148387753977 18

88378198587077823426167764978547808756246115090591034324707146225186949 52.415.240096429449348928530894810398630248865816766662999 5

39843046788666050160638129156997085.416.388484183426537766350753518180972865642314621443875164454555847769482631352751675692037.41

7.628580612875886694881648328579603114508131388106874704297602636435532791990880832689122.418.10170647963024244612324018467605759 80

150446009550749868752158484205015423343632508381159.419.164564540917831115611405017534017909465857739765762457304976112064054821533

4513341070281.420.2662710205480735617346452022100755074809023407208374441801919604845563638678145849451440.421.43083556146590467734

60502197440934169467600804865999014851680725486111854012659190521721.422.697106582013978239080695421954168924427662421207437345665

3600330331675492690805039973161.423.11279421434798829164267456416982623413744225016940372471505281055817787346703464230494882.424.1

8250487254938611555074410636524312658020849229014745928158881386149462839394269270468043.425.2952990868973744071934186705350693607

1765074245955118399664162441967250186097733500962925.426.4778039594467605227441627769003124872978592347496986432782304382811671302

5492002771430968.427.77310304634413492993758144743538184801550997720924982727487206270083963211589736272393893.428.1250907005790895

452681744224335694335313369211958948470553102500982006762370817390438248861.429.202401005213503038261932567177107618332887918916819

829782797456368284639448671475316218754.430.32749170579259258353010698961067705186422484011271467683810770646648531568575321436004

3615.431.529892711006095621792039556787784670197112759029534506620905162834769955134424689676262369.432.857384416798688205322146546

398461722061337599142249183459012869301255270820177904036305984.433.13872771278047838271141861031862463922584503581717836900799180

32136025225954602593712568353.434.2244661544603472032436332649584708114319787957314032873538930901437280496774780497748874337.435.3

631938672408255859550518752770954506578238315485816563618848933573305722729383091461442690.436.58766002170117278919868514023556626

20898026272799849437157779835010586219504163589210317027.437.950853888941998375153737015512661712747626458828566600077662876858389

1942233546680671759717.438.15385139106431711643524221557482279748374290861085515437934408603594478161737710269882076744.439.2489367

7995851695395061591712608896875850555449371181438711037372178370103971256950553836461.440.4027881710228340703858581327009117662422

4846310456696876645445975772848265708967220435913205.441.65172495098135102433647404982700073500075401759827878315356483334795121836

9680224170989749666.442.105451312200418509472233218252791250124300248070284575192001929323724066635389191391425662871.443.170623807

298553611905880623235491323624375649830112453507358412671675285005069415562415412537.444.27607511949897212137811384148828257374867

5897900397028699360341995399351640458606953841075408.445.4466989267975257332839944647237738973730515477305094822067187546670746366

45528022516256487945.446.722774046296497854662108306212056471121727445630906510906079096662473988285986629470097563353.447.11694729

73094023587946102770935830368494778993361415993112797851329548624931514651986354051298.448.189224701939052144260821107714788683961

6506438992322504018876947992022613217501281456451614651.449.3061719992484545030554313848083717208111285432353738497131674799321571

238149015933442805665949.450.4953967011875066473162524925231604047727791871346061001150551747313593851366517214899257280600.451.801

5687004359611503716838773315321255839077303699799498282226546635165089515533148342062946549.452.1296965401623467797687936369854692

5303566869175045860499432778293948758940882050363241320227149.453.2098534102059428948059620247186224655940594647874565999771500484

0583924030397583511583383173698.454.3395499503682896745747556617040917186297281565379152049714778313453268297127963387482470340084

7.455.54940336057423256938071768642271418422378762132537180494862787975116607001677217386408086574545.456.8889533109425222439554733

4812680590285351577786328700992010571109649289972956851261232789975392.457.1438356671516754813336191034549520087077303399188658814

86873359084765896974634068647640876549937.458.2327309982459277057291664382676325989930819177051945824788839301944151869475909199O8

873666525329.459.376566665397603187062785541722584607700812257624060463965757289279181083922224988556514543075266.460.6092976636435

30892791951979990217206693894175329255046444641219473596270869815908465388209600595.461.9858643290411340798547375217128018143947O6

432953315510410398508752777354792040897021902752675861.462.15951619926846649726466895017030190210886006082825705568550397282263736

25661856805487290962276456.463.2581026321725799052501427023415820835483307041235886067265438236979150980453897702509193714952317.4

64.41761883144104640251481165251188398565719076495184566241204779652055524606115754507996484677228773.465.67572146361362630776495435

48534660692055214690754342691385916202184675586569652210505678392181090.466.109334029505467271027976600736535005486271223402727993

15506394167390200192685406718502163069409863.467.176906175866629901804472036221881612406823370310271420068923103695748757792550589

29007841461590953.468.28624020537229717283244863695841661789309459371299994132239870453696507597194046564751000453100081 6.469.463146

3812391270746369206731802982302999179640232708332929101490653995175119552457651784599259176 9.470.74938658661142424746936931 0138714

8481930125577362702465168971944350502772313599022402785052359258 5.471.12125329678505513221062899833190130784929305217595410798098 0

7343500449794743315148005456965161843 54.472.196191955446197556957565929345772792668594307949581132632670453793550007197467505024 57

3547039776939.473.317445252231252689168194927677674100517887360125535240613651188143594986671799019825119243555961293.474.513637207

67745024612576085702344689318648166807511637324632164193714499386926652484969279059573823 2.475.83108245990870293529395578470112099

37043690282006516138599728300807399805410655446748120341516995 25.476.1344719667586153181419716641724567886890850696275767987106294

47201788497441033206952450482474743775 7.477.21758021274948561167136724264256888805952197244764196009662673020986249549513976141993

16858899137282.478.35205217950810092981333890681502567674860704207521875880725617741165099293617296837238216836465750 39.479.5696323

9225758654148470614945759456480812901452286071890388290762151348843131272979231385425457 12321.480.9216845717656874712980450562726 2

024155673605659807947771113908503316448136748569816469602261922873 60.481.149131696402327401278275120573021480636486507112094019661

5021992654677969798798427957009876873799968 1.482.24130015357889614840807962620028350479216011277190196743261610776878424511 6628412

61217058994930287041.483.39043184998122354968635474677330498542864661988399598709411830703425204209650825540787157763668286722.484.

6317320035601196980944343729735884902208067326558979545267344148030362872131366680200421675859857 3763.485.10221638535413432477807 8

911974689347564945335253989394162085272183728832930964492342791374522266860485.486.16538958571014629458752234927204819658702600851

957918961475871366403246165227815914479559128086543424 8.487.2676059710642806193656012612467375441519713437735685837768439858477612

945832426514875869658031322947 33.488.432995556774426913953123610518785740738997352293147773391602699511793756235520810632382557083

997728981.489.7006015278387075333187248717655232848909686960667163571684466853595550508187634621199695228871300237 14.490.1133597084

6131344472718484822843090256299660483598641305600493848713488070542842727523520799711277 52695.491.18341986124518419805905733540498

3231052093474442658048772849607023090385787304773487232160285825777 6409.492.29677956970649764278624218363341413361509007927864446 1

828854545510225266492733200762467368282938552910 4.493.48019943095168184084529951903839736466718355372130251060170415253331565228 00

37974249699528568764330551 3.494.776979000658179483631541702671811498282273632999946972430558698043540918772771175012166896851702 88

34617.495.125717843160986132447684122171020886294945718672124948303226285057685657105280914926186642542046721401 30.496.203415743226

804080810838292438202036123173081972119645546282154862039748982558032427403332227217009747 47.497.3291335863877902132585224146092 22

922418118800644244593849508439919725406087838947353589974769263731148 77.498.53254932961459429406936070704742495854129188261636423 9

3957905947817651550703969797809933069964807408962 4.499.86168291600238450732788312165664788095941068326060883324529903470149056115 8

235927134583281765744472045 01.500.1394232245616978801397243828704072839500702565876973072641089629483255716228632906915576588762 22

521294125.501.2255915161619363308725126950360720720460113249137581905886388664184746277386868834050159870527969684986 26.502.3650147

40723634211012237077906479355996081581501455497852747829366800199361550174096573645929019489792751.503.590606256885570541884749772

94255142804209290641521368844138669578527482710023705750158963298181645829137 7.504.9556209976092047528969868508490307840381744879 1

66691862941345251520750264617872315981632789108359480841 28.505.15462272544947752947817366237915822120802673943318828747355212209 37

34985356202428909975291189265240637550 5.506.250184825210398004767872347464061299611844188224855206102965574608942488002381152069 79

16190803488354459633.507.4048075506598755342460460098432195208198709276580434935765176967026774733585835809797669102696140760835 13

8.508.65499237587027353901391835730728082043171511588289869967948327131161996136096473304955852934996291152947 71.509.10597999265301

490732599643671505003412515860435409421932560009680142974347195483140293254396195769876129909.510.171479230240042261227388272445 77

8116168330115942384089293548423932591739608051304707888396896953989914246 80.511.277459222893057168553384709160828150293488720296 47

8308619148520734021483080006136110820940858911688675545 89.512.44893845313309942978077298160660626646181883623886239791269694466661

3222688057440818709337755865678589792 69.513.72639767602615659833415769076743441675530755653534070653184546540063470576806357692953

02786147773672653385 8.514.11753361291592560281149306723740406832171263927742031044445424100672479284561210177482396163706430458551

3127.515.19017338051854126264490883631414750999724339493095438109763878754678826342241845946777698949854204131204 6985.516.307706993

434466865456401903551551578318956034208374691542093028553513056268030561242600951135606345897560112.517.4978803739530081281013107 3

98656990883161994291393290726397318161003013196904490207103779406341483872096070 97.518.805587367387474993557712643417250666635155 4

6334770376418182484465381437595847958195297889176975473310716720 9.519.130346774134048312165902338328294975495135489248703283682155

666075411569564892860266335683240390312031677430 6.520.21090551087279581152167360267002004215865103558347366010033815054079300716 07

408184616335724173657853423941515.521.341252285006844123687575940998315017653786524832176943782493816616204576725633678727969255 65

77560973740715821.522.5521577958796399352092495436683350598124375604156506038828319671569975838863744971896028280751218827164657 33

6.523.8934100808864840588968254846666500774662240852478275476653257837732021606120081759175720837328779800905373157.524.14455678767

66123994106075028334985137278661645663478151548157750930199744498382673107174911807999862807003049 3.525.23389779576526080530029005

130016352147448857309113056992134835347034019051103908490247469955408778428975403650.526.37845458344187320471089755413366203520235

473765747838507616412856336016496087735221319219073488777057045434143.527.61235237920713401001118760543382555667684331074860895499

751248203370035547191643711566689028897555486020837793.528.99080696264900721472208515956748759187919804840608734007367661059706052

043279378932885908102386332543066271936.529.16031593418561412247332727650013131485560413591546962950711890926307608759047102264445

2597131283888029087109729.530.2593966304505148439455357924568800740435239407560783635144865703227821396337504015773385052336702205

72153381665.531.419712564636128966418863068957011388899128076671547993021605479585858227224221424221791102364954108601240491394.532

.679109195086643810364398861413891462942652017427626356536092049908640366857971825799129607598624329173393873059.533.1098821759722

77277678326193037090285184178009409917434955769752949449859408219325002092070996357843774634364453.534.17779309548094165871476607

91784794314784432111526800706093789579403138960940165075820050317562202766948028237512.535.287675271453218936393092272215569716662

6212205625975055651487108897637555022358325840971027525781204722662601965.536.4654683669341605951078583513940491481410644317152775

761745276688300776515962523401661021345087983971670690839477.537.75314363838737953150095062360961886480368565227787508173967637971

98414070984881727501992372613765176393353441442.538.121861200532154012660880897500366801294475008399315265791420404854991905869474

051291630137177017491480640 44.280919.539.197175564370891965810975959861328687774843573627102773965388042826976046579322868566650 06

090315514324457397.722361.540.319036764903045978471856857361695489069318582026418039756808447681967952448796919858280198080172634 7

2521442.003280.541.516212329273937944282832817223024176844162155653520813722196490508943999028119788424930258983327777969788 39.7256

41.542.835249094176983922754689674584719665913480737679938853479004938190911951476916708283210457063500412695002 81.728921.543.135146

142345092186703752249180774384275764289333459667201201428699855950505036496708140716046828190664791 2.1454562.544.2186710517627905

789792212166392463508671123631013398520680206366890767901981953204991351173110328603359794 0.3183483.545.3538171941078827656829734 65

82002073514287665243468581878814077955906238524869897016994918891571567940245852.4638045.546.572488245870673344662194682459267086 0

099890155360256708561614162481391754468942906690843062267485397384379 2.7821528.547.926305439978556110345168148279287821152865667970

711489644302195807201560695593260839033495142464219140896 45.2459573.548.1498793685849229455007362830738554907162854683506737160500 4

636120553407361424875515081178013692127588793343.80281101.549.242509912582778556535253097901784272831572035147744865014476580786254

229683808081234715129651167697802023 08.32740674.550.392389281167701502035989380975639763547857503498418581064522941991788303298056 8

363855269097880889736899565 2.13021775.551.634899193750480058571242478877424036379429538646163446078999522778042532981864917620242 0

39439256671491979960.45762449.552.1027288474918181560607231859853063799927287042144582027143522464769830836279921754005768949227345

6451819361.258784224.553.1662187668668661619178474338730487836306716580790745473222521987547873369261786671626010988666660231667391

57.304546673.554.2689476143586843179785706198583551636234003622935327500366044452317704205541708425631779937893947961855851 8.563330

897.555.4351663812255504798964180537314039472540720203726072973588566439865577574803495097257790926560550278529767 5.867877570.556.7

041139955842347978749886735897591108774723826661400473954610892183281780345203522889570864454498240385619 4.431208467.557.1139280376

809785277771406727321163058131544403038747344754317733204885935514869862014736179101504851891538 7.0299086037.558.184339437239402007

564639540091092216900901678570488739214977882242321411354939021430369326554695467593010 06.4730294504.559.29826747492038053534178021

282320852271405611887436347369040965556281000490642600763184294446484595278216393.5029380541.560.48260691215978254290641975291 43007

3961495779744485221290538753780513141626136502906221227101954142037517399.9759675045.561.7808743870801630782481999657375092 6232901

391631921568659579719336794142116779103669405521548438737315733793.4789055586.562.1263481299239945621154619718651810001943971713764

0678995011847311730728374291560657562674865039287935325119.34548730631.563.2044355686320108699402819684389319264272985630083 2835860

96981924541014258596947102450322701988316166689849 8.69337786217.564.3307836985560054320557439403041129266216957343847351485598166 65

5714087096026103168206590188492244960222361 8.03886516848.565.53521926718801630199602590874304485304899429739306350716951 4858025510

135462305027065691289048056112691221 16.73224303065.566.86600296574402173405176984904715777967069003177779865572933152359691884506 49

15343886350307897280608713457 34.77110819913.567.140122223293203803604779575779020263271968432917086216289884638162242898052722037 09

520415969453367214046785.150335122978.568.226722519867605977009956560633736041239037436094866081862817790521934782559213571 4838391

90484261733011 81358.627445942891.569.366844743160809780614736136462756304511005869011952298152702428684177680611935608579043350178 7

9540515228143.777781065869.570.593567263028415757624692697146492345750043305106818380015520219206112463171149180062882540663057 1381

6409502.405227008760.571.960412006189225538239428833609248650261049174118770678168222647890290143783084788641925890841852543316 3764

6.183008074629.572.155397926921764129586412153075574099601109247922558905818374286709640260695423396870480843150490968148047 14.8588

235083389.573.2514391275406866834103550364364989646272141653344359736351965514986692750737318757346734322346762224796 8479.4771243 1

58018.574.40683705446245081299676718951207306422832341325699487945357083820830953576915527260515427538516719062773194.3359478241 40

7.575.658276182003137496407122225948572028855537578591430853088767389706978810842887148339827707619843413107 41673.8130721399425.576

.106511323646558830940388941546064509308386099184842573254233822791528834661204242094498198300501060373514 86.81490199640832.577.172

3389418468725805811011641409217121939398570439856585631105617622267157454929569284809690624854016842565 4.19620921040257.578.2788502

654934314115214901056869862215023259562288282318173443845537555504066971990229791673629864620577714 1.01111120681089.579.45118920734

030399210259126982790793369626581327281389038045494631598226615219015595146013642547186374202795.20732041721346.580.73003947283373

540362408137551489415519859176950164212219779933086973781655888735497443930378845832579979936.21843162402435.581.11812268017403939

572667264534280208889485758277445601257825427718572008271107751092589944021393018954182 73.142575204123781.582.19112681530077 7479935

0754020857696244093449352276098134776053608055457899269964865900338744002388515341626 6.764418366526216.583.30924968331818141950774

266662004983329883069350505541473543078852411779819810423768262381842163187048834539.906993570649997.584.500376498618958899442 81806

870581945770817562873266522821303614932966358812510072427265769282187072202250806.671411937176213.585.809626181937140318950560 73532

586929100700632223772064294846693785378138632320496195528151124350259251085346.578405507826210.586.131000268055609921839337880 40316

887487151819509703858711615030871834449744483056862279392040653733145333615.3249817445002423.587.21196288624932395373439395393 57558

039722188273208106514109970025037226360771510648183220715308875907044214 9.9828222952828633.588.342963154304933875573731834338924 67

884373702241784923852714731122206713352198163344111599193742492215775765.3078040397831056.589.55492604055425782930812578827468 04828

1595584973865988993814431372578976959913269825943806346831251286217915.2906263350659689.590.89788919485919170488185762261360516165

969287215650912846529162494785690312111433170055405540573743501993680.5984303748490745.591.1452815235413449534189983410888285644475

6487218951690184034359386736466727202470299599921188740499478821159.58890567099150434.592.23507044302726412390718410335018908061353

415940516781468687275636215035758413613616605461742797873829020527.64874870847641179.593.380351966568609077326182444439017645061099

03159468471652721635022951502485616083916205382931538373307841687.23765437946791613.594.615422409595873201233366547789206725674633

19099985253121408910659166538244029697532810844674336247136862214.88640308794432792.595.995774376164482278559548992228224370735732

22259453724774130545682118040729645781449016227605874620444703902.12405746741224405.596.1611196785760355479792915540017431096410365

4135943897789553945634128457897367547898182707228021086758156611.701046055535657197.597.2606971161924837758352464532245655467146097

6361889270266967000202340261970332126043084329988608548802627001.913451802276881602.598.421816794768519323814538007226308656355646

30497833168056520945836468719867699673941267037216629635560783613.614497357812538799.599.68251391096100309964978446045087420307025

606859722438323487946038808981838031799984351367205238184363410615.527949660089420401.600.11043307057295224234643224676771828594259

023735755560638000889187527770170573147392561840442186781992419422.9142447517901959200.601.1786844616690525523114106928128057062496

1584421727804470349683791408668354376327390996977162710600428760484.467039717799137960 1.602.2891175322420047946578429395805239921 92

20608157483365108350572978936438524949474783558817604897382421179907.381284469589333880 1.603.467801993911057346969253632393329698 44

18219257921116957870025677034510687932580217455579476760798284994039 1.8483241873884718402.604.7569195261530621416270965719738536906

340280073669453468705082974928154540427527695811461237250536527112029 9.2296086569778057203.605.122472152006411948859635020436718338

907584993315905704265751086519626652283601079132670407140113348121060 69.1077932844366277560 5.606.1981641046217181630223446776341037

079709877940526002389528019162689081976878763560907850195126187133921809 9.0307541501344083280 8.607.3206362566281301118819796980708 2

204687857278736850594321855300278853484997147743522345542665273206151324168.13854743457103608413.608.51880036124984827490432437570

4925754849560581421106182171354919057443047659353791314240446165350774905422 67.1693015847054444122 1.609.83943661787797838678630407 3

77574780172813336878961212538990792184597789763083122653769587281808283641866435.30784901927648049634.610.13582369791278266616906 2

844948067355657769395021071830756126284090342094529018501785193631898343361132408 70.2477150603981924908 55.611.21976735970058050484 7

693252325642135830582731900033043295117076274939884292101624438963219180151644774275 13.7784999623258405404 89.612.355591057613363171

016756097273709491488352126921104874051243360365281978821120126224156851078495005906683 84.0262150227240330313 44.613.575358417313943

67586444934959935162731893485882113791734636043664022186311322175066312007025864665068095897.80471498504987357 1833.614.930949474927

306846881205446873061118807286985742242791397603797005503841934341876887276921337141656587 64281.8309300077739066031 77.615.150630789

22412505227456547964724127461262218445633807087439642336457257050475636275503969915957883072686017.9635644992823780175010.616.2437

2573671685573696268602433454738649335088303056235001415680306512295469819055044376739129329299638562446.1466575000597686778187.617

.3943565259409807892372515039817886611059730674869004208885532264296955252029469131988070904528718271124 8464.1102219993421466953 19

7.618.638082262657836526199937528316336047599323950517462770902710029494818479901137463642574481746164823498 10910.25687949940191537

31384.619.1032438788598817315437189032298124708705297018004363191791263255924514005104084376841381572199036650610 5937.436710149874

40620684581.620.167052105125665384163712656061446075630462096852182596269397328541933248500522184048395605394520 14741087028.462398

09981459774415965.621.2702959839855471157074315592912585465009917986526189154485236541343846490109306217325337626 1442381247192965.8

99108249689003951005 46.622.437348089111212499871144215352704622131453895504801511717920982676317897511452805780 92936800894395988279

994.36150634950360169516511.623.707644073096759615578575774643963168632445694157420427166444636810702546522383427 51346313062336777 2

35472960.260614599192605646 17057.624.11449921622079721154497199899966677907638995896622219388843656194870204440338 3623329439249863 2

3117322375295.462212094869620734133568.625.1852636235304731731028295764640630959396345283819642366050810256297722990556219660807855

6292556795045922591.488273554788881298750625.626.299762839751270384647801575463729875016024487348186430493517587578474343459005589

41022481278879912368297886.950485649658502032884193.627.485026463281743557750631151927792970955659015730150667098598613208246642514

62755549101037571436707414220478.438759204447383331634818.628.784789303033013942398432727391522845971683503078337097592116200786720

98597363314490123518850316619782518365.389244854105885364519011.629.126981576631475750014906387931931581692734251880848776469071481

39949676284882607003922455642175332719673884.3828004058553268696153829.630.2054605069347771442547496606710838662899026021886824862

2828310147816886144618938452934807527206994697925720.9217248912659154060672840.631.33244208356625289426965604860301544798263685406

953126269735458287766562429501545456857263169382327417599605.3045252971212422756826669.632.537902590501030038524405709274099314272

53945625821374892563768435583448574120483909792070696589322115525326.2262501883871576817499509.633.87034467406728293279406175787711

476225517631032774501162299226723350011003622029366649333865971649533124931.5307754855083999574326178.634.1408247264568312971318467

4671512140765277157665859587605486299515893345957774251327644140456256097164865025.77570256738955576391825687.635.2278591938635595

9041125292250283288387828920769137037721716222188228347058136454264309073842853262118177518.92878011594039575966151865.636.36866392

032039088754309966921795429153106078434996625327202521704121693015910705591953214299109359283042544.70448268332995152357977552.637

.59654311418395047795435259172078717540934999204133663048918743892350040074047159856262288141962621401220063.633262799270347283241

29417.638.96522703450434136549745226093874146694041077639130288376121265596471733089957865448215502441071980684262608.337745482600

29880682106969.639.15617701486882918434518048526595286423497607684326395142504000948882177316400502530447779058303460208548267.197

100828187064609006236386.640.25269971831926332089492571135982701092901715448239423980116127508529350625396289075269329302410658276

974528.030875376447094489688343355.641.40887673318809250524010619662577987516399323132565819122620128457411527941796791605717108360

714118485522795.227976204634159098694579741.642.66157645150735582613503190798560688609301038580805243102736255965940878567193080680

986437663124776762497323.258851581081253588382923096.643.10704531846954483313751381046113867612570036171337106222535638442335240650

898987228670354602383889524802011.8486827785715412687077502837.644.1732029636202804157510170012596993647350014002941763053280926403

8929328507618295296768998368696367201051744.1745679366796666275460425933.645.280248282089825248888530811720838040860701762007547367

55344902481264569158517282525439352971080256725853756.0232507152512078962537928770.646.4534512457101056646395478129805374055957031

6230172367288154166520193897666135577822208351339776623926905500.1973186519308745237998354703.647.73369952779993091352807862470137

544645640492430927104043499069001458466824652860347647704310856880652759256.2210693671820824200536283473.648.11871507735100365781676264376819128520521080866109947133165323552165236449078843816985605565063350457966475.641888801911295694385346381 76.649.19208503013

0996749169570506238328829850851301092026575375152304523110831315441298517503759961490385232424O1.2639957386295039363907092l649.650.

310800107482000406986333150006520115056062109753126046706805540044763195806229736687359815612123889812O8876.905884540540799630776O

5559825.651.50288513761299715615590365624484894490691341084515262208195784456787402712167103520486357557361427504451278.1698802791

7030356716676481474.652.813685245094997563142236806251369059962975520598278668788763384612637222927900771892223391185738164856601 5

5.0757648197111031979428204 1299.653.13165703827079947192981404624962180048698889314434312908707212291805112500495718070970869667593

524399011143.32456450988814067651095852277 3.654.21302556278029922824403772687475870648328644520417099596594846137931484729774725789

89310357945090604757715 8.83214099185925099630524056407 2.655.34468260105109870017385177312438050697027533834851412505302058429736 59

72302704438608639732470444304465883 02.156705501747391672816199086845.656.5577081638313979284178894999991392134535617835526851210 18

96904567668081960045169650757076826495336494165460.98884649360664266912143965091 7.657.90239076488249662859174127312351972042383712

190119924607198962997404679190315613511621050073539766940753763.145551995354034341937638737762.658.1460098928713894557009630773 1226

5893387739890545388436709095867565072761150360783162378126900035103434919 22.4134398488960677011059078388679.659.2362489693596391185

601372046246178654301236027355083613162948305624774403406763966739991769735748703756 7298.727995048431471135299671712 6441.660.382258

862231028574261100281936883758817863493280896798025390698127550201491037179836377303873609973810 59221.141434897327538836405579551 5

120.661.618507831590667692821237486561501624247987096016405159341685528690027641831713576510376480847184844186 26519.86942994575900

997170525126415 61.662.100076669382169626708233776849838538306585058929730195736707622681757784332275075634675378472079481799685 74.

101086484308654880811083081566 81.663.161927452541236395990357525505988700731383768531370711670876175550760548515446433285713026556

79796621831226.088029478884555877981608207982 42.664.26200412192340602269859130235582723903796882746110090740758379823251833284772 15

089203884050288774480179980 0.189115963193210758792691289549 23.665.4239315744646424186889488278618159397693525959924716190784599737 8

327888136316794220610143158567541423631026.27714544207776663677429949753165.666.685935696388048441387540130217643178807321423453 57

252648604377201579721421088945112648983661455286225430826 466261405270977395566990787080 88.667.110986727085269086007648895807945911

857667401944604414556450374579907609557405739333259126820022827649061 85.274340684734874403234129028461253.668.17958029672407393014

640290882971022973839954428996166720505475178148733097849468444590811048147811387449267.920966825261972142790828107169341.669.29056

7023809343016154051804637656141596066946234566081761505126361394940535900423779167237301500941523554 53.1953075099968465460249571 3

5630594.670.4701473205334169463004547134673663713344664905245277489665598781428822715143951082250753477829790553980 4721.11627433525

881868881578524279993 5.671.760714344342759962454506518105022512930533436759093830728065004504277212050295532004242585084479996921 6

0174.311581845255665234840742378430529.672.123086166487617690875496123157238888426499992728362157969462488264715948356469064022931 7

932867459052319648 9.542785618051448392365652762123046 4.673.1991576009218936871209467749677411397195533364042715410422689887151436 69

5614986172233560517951939049241250 6.973943802577014915849726999966099 3.674.3222437674095113779964428981249800281460533291326336990 1

1731476979859617917967681246287845081939810156089 96.516729420628463308215379762089145 7.675.521401368331405065117389673092721167865 6

066655369052400540004656950032874794662984696438968771337150802150 3.490673223205478224065106762055245 0.676.8436451357409164431138 32

5712177011960116599946695389390657319426748629053974339797159317419590735252363050 0.00740264383394153228048652414439 07.677.13650465

0407232150823122224431042236387726666020644417911973240836986619287690027818557563883620724031652 00.349807586703941975634559328619

96357.678.22086916398132379513450548155281235598889266548759831181854643510447290982743342579015073807952807655528250.3505478510873

361288626079810344026 4.679.3573738143885559459576277059838545923766193315082427297305196759414595291151234536087083019631488005869

3450.70035543779127810449716730965436621.680.578242978369879741092133187536666948365511996995841041549066111045932438942556879 39885

904004267687714221701.05090328887861423335977529068876885.681.93561679275843568704976089352052154074213132850408377127958578698739

196805768033300756734200582567772915151.75125872666989233785694260034313506.682.151385977112831542814189408105718848910764332549992

48128286518980333244070002372124064263820485025548713685.280216201554850657121671789103190391.683.244947656388675111519165497457771

00298497746540040085841082376850207163750579175454139937240543282326005200.455342074221839890907366049137503897.684.39633363350150

665433335490556348985189574179795039333969368895830540407820581547578204201061028307874718885.735558275776690548029037838240 69428

8.685.64128128989018176585252040302126085488071926335079419810451272680747571571160723032344138301571590200724086.19090034999853043

8936403887378198185.686.103761492339168842018587530858475070677646106130118753779820168511287979391742270610548339362599898075 4429

7.1926458625775220986965441725618892473.687.167889621328187018603839571160601156165718032465198173590271441192035550962902993642892

47766417148827616705.8117358975773751425901845612997090658.688.27165111366735586062242710201907622684336413859531692737009160970332

353035464526425344081702677138635161003.0043817601548972412867287338615983131.689.439540734995542879226266673179677383009082171060

5151009603630508953590813175482578963332946909428746277770 8.8161176577322723838769132951613073789.690.71119184866289873984869377519

87536098524463096558320283304546605986826116721935221497741117177142609793871 1.8204994178871696251636420290229056920.691.1150732583

65844161907496044837843099286152848071634712929081771149404169298974178004610740640865713560 71642.0636617075619442009040555324184 2

130709.692.18619244323213403589236542235771846027139747903721791576212723720927243046619353021958815181263713965865513.245711649350

661163420419735320711876 29.693.30126570159797819779986146719556155955755032710885262869120900835867659976516770822419889245350285 3

21937155.309373356912605364324475267739133183 38.694.48745814483011223369222688955328001982894780614607054445333624556794903023136 12

38443787044266139992878026 68.5550850062632665277448950030598450596 7.695.78872384642809043149208835674884157938649813325492317314 45

452539266256299965289466679859367196428460973982 3.8644583631758718920693702707989782430 5.696.1276181991258202665184315246302121599

2154459394009937175978814994945746602278901851117729809857828389754249.241954336943913841981426527385882330272.697.2064905837686293

0966764036030509631786019440726559168907424267534212002902244191317797589177054256850728231.628400173261501031188363554465780154 57

7.698.33410878289444957618607188493530847778173900120569106083403082529157749504523093168915318986912085240482480.870354510205414 8

731697900818516624848 49.699.540599366663078885853712245240404795641933408471282749908273500633697524067672844867129081639663420912

10712.4987546834669159043581536363174426394 26.700.874708149557528462039784130175713273423672409676973810742304325925275019112903776

5562822715087842733169319 3.369109193672330777527943718169105124275.701.14153075162206073478934963754161180690656058181482565606505 7

78265589725431805766214234113531484476942290390.58678638771392466818860973544865477637 01.702.229001566577813580993328050559183134 2

89278227825230371392882152484247562293480397979693624657231967545970 9.923697307081157745941404107265565288797 6.703.370532318199874 3

15782677688100794941155488404597348693204345997904322010547405701940310497780567966177501 00.510483694795082414130013842714220065 13

77.704.59953388477768789677600573865997807540441622737987173034363421315274676677675374173827986024629116293209810.434181001876240 1

600714179499797853539653.705.9700662029775622125586834267607730165599046319772204235479802110570687773241594436785903580268591291 0

959910.944664696671322574201431792694005419133 0.706.15696000877552501093346891654207510919643208593570921538916144242098155441009

3185416870218273150292041697 2.1378845698547562734272849742673790773098 3.707.2539666290732812321893372592181524108524225491334312 57

74395946352668843214250726290954605763000094211512963.23235103952188853084742815353677961922313.708.410926637848806243122806175760 2

2752004885463506914047313312090594766998655259858145123307945731597131929935.3702356093766448042747131278041586965329 6.709.664893 2

66922087475312143434978379930901277184202571730877080369474358418695105844360779137087316913434428 98.60258664889853333512214128 3

4093831575609.710.1075819904770893718434949610738607450950131819271712204010201275422028405247704425812012216544632884753728 3.39 2

82225827517813939685440914509701228905.711.174071317169298119374709304571698738185140900347428393488728164489638682394281027017279

1353631949798188157 3.2575408907173711474518995690486035328045 14.712.2816533076463874912182042656455594832801540822745996138897 48292

0318415229190514695984803570176582682941 8856.65482311654488896139158500996311323403341 9.713.4557246248156856105929135702172582214 65

294982622028007378476456521480205313332496615759492380853248113004 29.91236400726226010884348457901171676683 7933.714.737377932462073

1018111178358628177047454490648966276212682247485533217282323839662142398493985115164071 9286.56718712380714907023506958897 48300008

71352.715.1193102557277758712404031406080075926210744047518655628646701205074801933545716462829999341779364764520197 1.6479551131069

40917907855416798654676770928 5.716.193048048973983181421514924194289363095619311241528324991492595362812366177810042904423919117 78

76280927390 0.304673825487655824931362375696137676858063 7.717.31235830470175905266191806480229695571669371599339388785616271587 0292

55953238168918742385329572410454475871.95262893859459674283921779249479235362899 22.718.5054063536757422340834329889965863188123130

2723492221284765531123310492571019173209184777241351173263749772.25730276408225256777058016819093003048705 59.719.817764658377501286

745351053798883274529006743228316100703818027103397485242573421279271625709235837182256 44.20993170267684931060979796068572238411 60

481.720.132317101205324352062878404279546959334131977046323831355147333833650241095276515337111939812274756981975 41.646723446675910

18783803781288766524146031040.721.214093567043074480757413509659435286787032651369155441425529136543989989619533857465039102383198

34070020106.06771661694359511889901760895623747987191521.722.3464106682483988328402919139389822461211646284154792727806764703776402

30714810372802151042195473097682176 47.71444006361950530673705542184390272133222561.723.5605042352914733135977054235984175329081972

7978463471420620560692163022033434423026719014457867143838237753.78215668056310042563607303080014020120414082.724.9069149035398721

46437997337537399779029361908200113986986882077299270451049154603069341186774144536064554 01.4965967441826057323731284526440429 2253

636643.725.14674191388313454600357027611358173119375591879847487011930876842209006713834988333365313313528159744469315.527875342474

570615800920148344418312374050725.726.23743340423712176064737000986732170909669210961848626881799697615201711224326534364058725181

269605105114855.677535016892831189038232993608822604627687368.727.38417531812025630665094028598090344029044802841696113893730574457

41071793816152269742403849479776484958417 1.20541035936740180483915314195324091700173809 3.728.6216087223573780672983102958482251493

87140138035447407755302720726124291624880570614827636760673699546990 26.882945376260232993877386135562063521629425461.729.100578404

0477634373949250581829128589677588166452408546692608465300231471006495797589068021708651348042831 9.8088355735627634798716539277 51

5304438631163554.730.16273927628350124412475608776773537390647283044878559544479111860263557626313763682038956584693250475898222.49

713011118878677925939254130773679602605890 15.731.26331768033126468151968114595064823287423164709402645011405196513265872336378721 65

79296368017797639563265 42.30596568475155025913104646905926723988917525 69.732.42605695661476592564443723371838360678070447754281204

5558843083735294299626924853399685933864730144322247 64.803095795940337038390439010367004035915234158 4.733.6893746369460306071641183

79669031839654936124636838495672895048867953022990712069978982301882527783885513 07.109061480691887297521485479426271275804409415 3.

734.1115431593560796532808555613387415446435640602179650541231738132603247322617636923378668235747257928207760 7.191215727663222433

5911924489793275311719643573 7.735.180480623050682713997267399305644728609057672681648903690463318147120034560834899335765053762978

5712093273 7.902121875732411163343340996921954658752405298 90.736.29202378240676236727812296064438627325262173289961395781363713140 74

4476682259859167363187733770436403010345.093337603395633596934533445901282189924369656 27.737.47250440545744508127539035995003100 18

61679405581262861504100449554564801383433491009396931100682935239430 82.995459479128044760277874442823236848676774955 17.738.76452818

7864207448553513320594417275114301138480876819317737580962009568206032082683028808438387299269534 28.0887970825236783572124078887 24

51903860114461144.739.123703259332165252982890368054444827697598054406213968082183803051657436958946557369242573953907023450896 51.1

0842565616517231174902823315477558872779195666 1.740.20015607811858599783824170011388655520902816825430165001395756114785839377954 9

76563754545479774575337784993.917305364417540147470269022027227492587906417805.741.323859337450751250821132068168331382906626222660

51561809614136419951583073849632300678802875165277682874645.025731020582712459219297255182003081315698374466.742.52401541556933724

865937376828221793811565439091481726811009892534737422451804608864433348354939853020659638.943036385000252606689566277209230573903

604792271.743.84787475302008849948050583645054932102228061357533288620624028954689005525654241165112151230105130703534283.968767405

582965065908863532391233655219303166737.744.13718901685894257481398796047327672591379350044901501543163392148942642797745885002954

549958504498372419392.2911803790583217672598429809600464229122907959008.745.2219764921609514247620385441183316580160215618065483040

5225795044411543350311309119465765081515011442772820.6880571196166182738507293341991697884342211125745.746.359165509019893999576026

50459160838392981506225556331948389187193354186148057194122420315040019509815192212.9792374986749400411105723151592162113465119084

753.747.58114200118084542433806504870994004194583662406211162353614982237765729498368503241886080121534521257965033.667294618291558

3149613016493583859997807330210498.748.9403075102007394239140915533015484258756516863176749430200416943111991564642569736430639516

1554031073157246.6465321169664983560718739645176022111272449295251.749.152144951138158484825215660201148846782148831037978656655619

15166888564514479420060619247528308855233112228.03138267352580566710331756138759882109079779505749.750.2461757021582324272166248155

313036893697139996697461509576233211000055607912198979704988704446425834042795 2.696035885222455502710504957839359042203522288 01000

.751.398320653296390912041840475732452536151862830707724807613242472768891205936014098576691345727731135735401 80.727418558748261169

813822519226957863294320083067 49.752.644496355454623339258465291263756225521576830377470958570865793868896766727233996547190216172

37371913968133.42345444397071667252432747706631690549784237107749.753.104281700875101425130030576699620876167343966108519576618410

826663778797266324809512388156190010485487508 31.415087300271897784233814999629327476879216245414498.754.168731336420563759055877 10

582599649871950164914626666724754974060506684739390482091671071778072478574014764 4.757432744668969451486247747335959167429000482 52

2247.755.273013037295665184185907682525617374886845615254786249093908232714447271205373018679495333997258342888984 76.1725200449408

672357200627469652866443082167279367 45.756.4417443737162289432417847883516138736063472644010529215694056387651157451444212278466 02

5118045062002904612 0.929952789609836687206310494301245811737217210458992.757.714757411011894127427692470877231248493192879655839170

663313871479563016349794246526097845801764543179445 97.102472834550703922926373241266532456045433938395737.758.115650178472812307066

9477259228845122099540144056892092232719510244678761494215474372700357606270743469907 1.803242562416054061013268373556777826778265 1

148854729.759.18712591957400171980971697301060763705927330237127312628960333817242417778440097208987982034080352866493531.51348984

58711244533059056976834310723828085087250466.760.302776098046814026876664698933492149269227316776962335512875289196892053933822519

5271498561014306030119260З.3167324082871785143191740712402088991610736236105195.761.48990201762081574668638167194409978632850061914

823546180247862736931623171822349161702967644223413167686134.8302222541583029676250797689236399715438821323355661.762.7926781156676

2977356304637087759193559772793592519779731535391656620828565204601114417953254366473468878738.14695466244548148194425384016384887

07049557559460856.763.12825801332884455202494280428216917219262285550734332591178325439355245173702695027612092089858988663656487.2

9771769166037844495693336090874888422488378882816517.764.2075258248956075293812474413699283657523956490998631056433186460501732803

0223155139053887415295636010544361.11241315790492659315135874492513377129537936442277373.765.33578383822445208140619024565209753794

501850460720643155510190044372573203925850166665979505154624674200848.41013084956530503810829210583388265552026315325093890.766.54

330966312005961078743768702202590369741415370706953719842054649389901234149005305719866920450260684745209.52254400747023163125965

085075901642681564251767371263.767.87909350134451169219362793267412344164243265831427596875352244693762474438074855472385846425604

885358946057.93267485703553666936794295659269908233590567092465153.768.14224031644645713029810656196961493453398468120213455059519

429934315237567222386077810571334605514604369126.745521886450576830062759380735191550915154818859836416.769.2301496665809082995174

693552370272786982279470335621474705465440369148501102987162504915597716600314026373253878937215413049699955367639448145914874538

5952301569.770.37238998302736542981557591720664221323221262823569669806574084338006722578252257702859727311771517744632859.28431125

8604707327062313057129673010063900204812137985.771.6025396496082737293330452724436694919304405752692588455362873874169820758928212

9327908883288937520884896591.823100630758837824061866733524154469212645590764439554.772.9749296326356391591486211896503117051626532

0350495554360202823079704930167534387030768610600709038629529451.107411889363545151124179790653827479276545795576577539.773.1577469

2822439128884816664620939811970930937787742143891383156182140313775681651635867749388964655951442604.29305125201223829751860465241

77981948489191386341017093.774.255239891487955204763028765174429290225574698227916993274034384901108067924350903389446104490355598

14395549.4037924409485928126310226314831809427765737181917594632.775.41298681971234649361119541138382740993488407610533843218786594

672251120568116741974812359838000215765838153.6968436929608311101496272839009791376254928568258611725.776.6682267112003016983742241

7655825670016045877433325542546190033162361927360551832313756970287035775580233703.1006361339094239227806499153841600804020665750

176206357.777.10812135309126481919854195879420841100953428504385938576497662783461304792866857428856933012503599134607185.679747982

68702550329302771992851392180275594318434818082.778.174944024211294989035964376450034081025580162477184928311166660996974975289220

40660232630041207176692630555.989811596077967895571092711466929929842962600686110244394.779.2830653773025598082345063352442424920351

1444752104431407614328883158802321788898089089563053710775827237741.66955957876499339886412043139544385164571854387045842521.780.4

5800940151385479727047071169427657306069460999822924238730994982856299850710938749322193094917952519868297.65937117484296129443521

314286237378148868114455656866960.781.74107477881641460550497704693851906509580905751927355646345323866015102172499836838411756148

628728347106039.32893075360795469329933357425781763313439968842702709481.782.119908418033026940277544775863279563815650366751750279

88507631884887140202321077558773394924354668086697433.698830192845091598773454671712019141462308083298359576441.783.19401589591466

840082804248055713147032523127250367763553142164271488650419571061242614570539217540921408037.63172326820588706810338802913780090

4775748052141062285922.784.3139243139476953411055872564204110341408816392554279154164979615637579062189213880138796546357220900810

5471.330553461050978666876842700849820046238056135439421862363.785.5079402098623637419336297369775425044661129117591055509479196042

7864441041463200044002536002789749929513508.962276729256865734980230729987620951013804187580484148285.786.821864523810059083039216

99339795353860699455101453346636441756584240231663355338845390501466361958937618980.292830190307844401857073430837440997251860323

019906010648.787.13298047336724228249728467303754960430731074627736390173123371701210467270481853888939303746915170886713248.92551

0691956471013683730416082506194826566451060039015893.788.215166925748248190801206372377344958168010201378817248367675473596344904

3681738777347835389355136678047514 6.954793710987255453869437759166250294551752483362029616958 1.789.348147399115490473298491 04541489

45624753209476561811500989091906084495770729924166241765764046653766718 8395.880304402943726467553168175248756489378318934422068632

8514.790.563314324863738664099697417792239520643331149034998398466584664204794481441166294358960115340179044476635 42.83509811393098

19214226059344150067839300714177840982498095.791.911461723979229137398188463207134083118652096691179548565493854813244058514158710

9831366917448444211485193 8.715402516874708388975774109663763273308390352206166882660 9.792.147477604884296780149788588 0999373603761 9

8324572617794703207851901803853995532500534209680708502346562515 48.155050063080569031039838004407877005723846176999026513247 04.793

.2386237772822196938896074344206507686880635342417357495597572373831282598469483716325233498829867886773674 2.02659031476803 9869937

415415374253333054685212219643201513 13.794.38610138216651647403939602252058812906426185881435354426296508928493211384248 0872166733

0305914891352398829 0.181640377848608900977253419782130338778531389218669714760 17.795.624725159448736167929003456941238897752325393

056089293822722326668060373689429243799256380474475923917250 32.208230692616648770914668835156383671833216601438312916273 30.796.101

0826541615252641968399479461827026816587251870442838085687415952992487531910115965989411065965059157133 2.23898710704652576718919 22

2549385140106117479906569826310334 7.797.163555170106398880989740293640306592456891264492653213190840974262105286122133935976524579

15404409830743835.45981017630819064428065910900948976824449645920952955473067 7.798.264637824267924145186580241586489295138549989 67

969749699940971585740453487532494757312352026064060422315167.6987972833547164114698513345033411693056712582752278178340 24.799.42819

2994374323026176320535226795887595441254172350710190250690119509820997458883549648099414684702530590 03.1586074596629070557505104 43

51283093755016771748475737256470 1.800.69283081864224717136290077681328518273399124385204820718966040597691435587278383112277161967

532530675374170.857404743017623467220361778016172106855838975759985190398725.801.11210238130165701975392213120400810703294324980243

98917379911096096424176870242714672419719090010009284331 7.40160122026805305229708722215290030444060066932447425629634 26.802.1813854

6316588173689021220888533662530634237418764471245695715020733385327430265457951913387653353160380734.487341694569815399019123399 9

5451751512618456690047277533621 51.803.2934878444675387566441343400893447323392856239900846041949482598169762709613269260467611057 8

553453253224051.888942914837868451316210622107417819566785236224947031632557 7.804.47487330763342049353434654897468135764562799817 7

729316651905410024310124235629580626280239662068064136047 86.37628460940768385033533402206193533469296980312541980696877 28.805.7683

61152100959250178480889064026089984913622167813920846853669841286395196956506673041345447602596668288 38.265227524245552301651544 64

41693531542597550393503668386013305.806.124323445973437974371282743803870744763054162034554323749875907986559651943258608729932158

51096706608043362.4641512133653236151986878666231288488952724842475786645570103 3.807.2011595611835338993891308327102733537615455242

5133571583456127497068829146295425939723629305572732574726246.2906739657898788453638423310400641643212479881826153484171433 8.808.3

25483007156971873760413576514144098524599686285890039584437182957247943406212868127168451566694391827696 08.7548251791552024605625

301976631930132165204724301940129741537 1.809.5266425683405057731495444092244174522861452105372257554189984579279362348691671275244

0474462242171757495855.045499144945081305926372528703257177537768460612809361391297 09.810.8521255754974776469099579857385615508107

4489682311579500343564088518417827537999565157319618911610940265463.8003243241002837664889027263664501907542889330430033743654508 0

.811.13787681438379834200595023949629790030968901073603415504224340988131204131445471231759779408115378269776131.884582346904536507

241527525506970736829205739365581273575674789.812.2230893719335461066969460380701540553907635004183457345425869739698304591419927 1

1882755113700006539363802678.264614779314564883890417798143615755904634632669881611012219869.813.360966186317344448702896277566451 95

5700452511154379889584830383851142500456447424200352907781219176335788 10.149197126219101391131945323650586492733840372035462884587

894658.814.584055558250890555399842315636606011091216011572725624127417357820972959598440136083108021481284569973814 88.413811905533

6662750223631217942022248638475004705344495600114527.815.945021744568235004102738593203057966791668522727105513712247741672115460 05

4887560283460929262503746309602985630090317527676661543084454447887413723153767408073801880091 85.816.152907730281912555950258090 8

8396639778828845342998311378396650994930884196533276963665689507437883162834178.697682093728643394117667156723899099001079038144 6

51875788123712.817.24740990473873605636053195020427219446745530570269366515519128411652038797082152566500298800062920625930208.553 9

82996903920160733098001268377973138310575818695925597613 2897.818.40031763502064861231079004108823859225574375913267677893915779406

58292299361542953016598830750080378876438 7.25166509063256355485076515799227707213938961396331111317642566 09.819.6477275397593846686

7132199129251078672319906483537044409434907818234961790697582096666237107563724414694595.8056480875364837155838631592606550452 77

0018978200703877403895 06.820.10480451747800332809821120323807493789789428239680472230335068722481788478431301162683227541506452820

345898.305731317816904727043462831725293211741708980374531815195046461 15.821.16957727145394179496534340236732601657021418888034176 6

71278559504305284657501059372349856252262825261815357.886296126570553098601849147651358716269478999352732519072450356 21.822.274381

788931945123063554605605400954468108471277146489016136282267870731359323605350330837937692780821612 56.19202744438745782564531197 93

76651928011187979727264334267496817 36.823.44395906038588691802889800797272697103832266015748825572892187731092357793433419907382 94

004603210334397661 4.07832357095801092424716112702801064428066697907999685333994717357.824.71834084931783204109245261357812792550643

1131434634744745058159578794309293657804424160238398013814261 37870.27035101534546874989247310640466257229185495880726118760744399 0

93.825.116229990970371895912135062155085489654475379159212300047398003688971788722799200349798963885833484770114 48.4348674586303479

6741396342334326732165725219378872580409473911 6450.826.18806407590215510002138032351289828220511849230267577452190381964685121 9652

16498079221498772563486619625235.4619025601648948424032107339837335788864376896694519228554835155 43.827.304294066872526995933515 38

56679837718595938714618880745693018233358230083749641811420139516114683509663668 3.89677001879524280981717415732700090054368988 3458

177726950222631993.828.49235814277468209595489570918088205406471236376456384909120564298267422802712916193422893933710321716261919

.3586725789601376522203848913107344794301275731276296498057061475 36.829.796652209647209091888411094848865825924306235226451923660 50

74663184972364020933430762428909485715681289860 3.255442597755380462037559048637735379973817456585807376755928779529.830.1289010352

4218911878433068040297478799890185989910157727517131093011714644292225050104718302856747852916052.2614115176715518114257943939 9484

698594039450297134370265616349270 65.831.2085662562069100279731717898878613705913324834217467696412220575619668700831315848086714 72

12342463534205912.5869557774470898576295502988586205239377762486299244403317563706594.832.3374672914490991467575024702908361585902

343433208483469163933684920840165260538353097186551519921138712196 4.848367295118641669055344692853467509878170751601268142987919 86

33659.833.546033547656009174730674260178697529181566826742595116557615426054050886609185420118390127275416749213278 77.4353230725657

3152668489499171208803381594700023119258331967623402 53.834.8835008391051083214881767304695336877718011700634434634740087945461 3490

3135239255428108782427408863084498 42.2836903676843731957402396845655555436941177518324607263075960973912.835.1429534386761117496218

850990648231216953367996806038580031624220600185789744424675546498909702825612297777 1.971901344025010472242513467627764357751 0064

752063653309627272331416 5.836.231303522586622581770702772111776490472516916686948204350563301514632069287966393097460769213023447 5

3822756.2002703807934477918165374360843199121204182503896114035934868428807 7.837.3742569612627343313925878711765996121678537163675 5

206235372572357465064826240886065211066018330600876800528.17217172481845826405905090371208426987142472559597673455621407602242.838.

605560483849356913163290643288376102640370633054500266704289025089282717550375253749571429396329456306232 84.3724421056119060558755

8833979640418199184297598558813814970091890319.839.979817445112091244555878514464975714808224349422052329058014748663933365812784 1

144016820895796354650742381 2.544613830430364319934639243508488451863267701581564872705914994925 61.840.1585377928961448157719169157

753351817448594982476552595762303773753216083363159368151253518975964921380470 9.691705593604227037581022758330489263385511067756 71

5301085561591382880.841.256519537407353940227504767221832753225681933189860492482031852241714944917594348255293560855560038645470

0.946166976647263469574486682681338108571837837914871788356153090875441.842.415057330303498755999421682997167934970541431437515752

05826222961703655325391028507041891275315653078351800.637872570251490507155509441011827371957348905671587089441714682258321.843.671

576867710852696226926450219000688196223364627376244540294081858751498171504633325712473608716569428988891.584039546898753976729996

123693165480529186743586458877797867773133762.844.10866341980143514522263481332161686231667647960648919965985563114757880514254149

183961313863618731002125069.2221912117150244483885505564704992852486535649258045967239582455392083.845.1758211065725204148453274583

4351693113629881606922682411388503933345395495969195517218438599705896696414958.38059516640489984606155016883981583330157223928445

04845037450228525845.846.284484526373955560067962271665133793452975295675716023773740670481032760102233447011797524633246276985400

27.6027863781199242944501007253103151185502258042102550812277032683917928.847.46030563294647597491328973000865072458927411174494284

788762570981448671506192540218398191063030524394954985.9833815445248241405116508941501309518517980434947055657314482912443773.848.7

4479015932043153498125200167378451804224940742065887166136638029551947516415884919577943526355152093495013.5861679226447484349617

516194604460704020238477049606469591515596361701.849.12050957922669075098945417316824352426315235191656017195489920901100061902260

842513797613458938567648844999.95695494671695725754734025136105770222538218911996662126905998508805474.850.194988595158733904487579

3733356219760673772926586260591210358470405525665390243100575540781157408258194501.3155717389814321010435154133071023092655845738

9046268596497514105167175.851.3154981743854246554770335465038655003305296445751862310759350560515531855616327351955302127051265050

7039501.27252668569838935859085566466816001149096676301042930723403512613972649.852.51048676954415855996461291983948747639790693723

381229019697090309210575210065704525308429082086733365234002.58809842467982145963437107797526232075655133690089199319901026719139

824.853.82598494392958321544164646634335297672843658180899852127290595914365893766228978044861450352599383872273503.860625110378210

81822522674264342233224751809991132130043304539333112473.854.133647171347374177540625938618284045312634351904281081146987686223576

46897629468257016987943468611723750750.644872353505803227785959782061868465300406943681221329363205566052252297.855.216245665740332

49908479058525261934298547801008518093327427828213794236274252366061503132978728550110978101.0309348645436243096084824563262106985

25158753672353459406510105385364770.856.349892837087706676625416523870903388298112361989462014421265968361518831718818343185201209

22197161834728851.675807218049427537394442238388079163825565697353574788769715671437617067.857.56613850282803917571020710912352273

128359037207464294769554425049946119446134200380023253900925711945706952.706742082593051847002924694714289862350724451025928248176

225776822981837.858.91603133991574585233562363299442611958170273406410496211681021886098002618016034698543374823122873780435804.382

5493006424793843973669331023690261762901483795030369459414482605989 04.859.14821698427437850280458307421179488508652931061387479098

12354469360441220641502350785666287240485857261427 5.70892913832355312314002916278166588885270145994054312851221672250835807 41.860.

2398201182659530880381454375112374970446995840202852871929164688221421246821662697771100035471714595065785 6.147184068387801061579 7

658560919027914703304747784934322068108673344179645.861.38803710254033159084272851172303238213122889463416007817415191575818624674

6316504855676632271220045232721 31.856113206711354184719795018873568680323031934719036560719027589842776038 6.862.6278572208062846788

808739492342698791759284786544453653670683845803283714284827746327866358183915047392998 8.0032972750991552462995608749654714717933

6240949752999292583845717719400 31.863.1015894323346616269723602460957302261307157373288605443541220300338514618174799279488463268 0

896115499720211.98594104818105094310193558938390401521163943442165665536448660470199700417.864.164375154415290094860447641019157214

04830858519430508089082886849188429896032820541212499039080030547113210.78627077569096646773189167688045116239097567537140965465 70

7045041971640448.865.265964586749951721832807887114887440179024322523165625244950898525735760777808133360971317199761460468334 22.77

22118238720174108338272662643551776026151097930663100215570551217134086 5.866.430339741165241816693255528134044654227332907717470 70

61357797670176200597381363387730963075905617659394663З.55848259956298387856571894314480633999359078516447596467862750554142981313.

867.696304327915193538526063415248932094406357230240636331380730665543355820515944472134067624790323226407800056.330694423435001289

39954620940916151759620589495754227470018456066314322178.868.112664406908043535521931894338297674863369013795810703751651043256097

58802540808109071639323808849923472668.988917702299798516796526515255396785758979668012201823937881206620457303491.869.18229483969

956288937453823586319088430400473681987433688972410981043317007700252830412315571712082187550674.621987144643298645736481136196312

937518600257507956051407899662686771625669.870.29495924660760642489647013020148855916737375061568504064137515306653075810241060939

48395489552093211102334З.610904846943097162533007651451709723277579925520157875345780869307228929160.871.477254086307169314271008З6

606467944347137848743555937753109926287696392817941313769896270467233014298574018.23289199158639580826948878764802266079618018302

8113926753680531994000554829.872.7722133329147757391674784962661680026387522380512444181724744159434946862818237470938022536275394

6409597361.843796838529492970802496439099732384073760108548271802099461401301229483989.873.1249467419221945053438486862330847446110

1307254868037957035736788204586144612368847927649582998696070817138.00766888301158887790719852267477550448699402915763857288531419

33295230038818.874.20216807521367207926059653585970154487488829635380482138760480947639533007430606318865672119274090711776874.1520

485668645381749874481665847487428943700400124657530952603334596459522807.875.32711481713586658460444522209278628948590136890248520

095796217735844119152042975166793321702272786782594012.1997174498761270528946466892595242473813640691701043259805745267891689561622

5.876.529282892349538663865041757952487834360789665256290022345566986834836521594735814856589938215468774943708863.917660167406652

2788209485584427299027573410918257007907583486024881490844328.77.8563977094854052484694869800452741238466910341587752233035291641 9

327771311516556652452315523819664276964898.5914834666167922807767415451037972376570981783526744050564093870379838646057.878.1385680

601834943912334528737997761958207480699415065245649096151028114234709901381381113093453665417713357 8.4983249483357457508658836400 9

480702279328322875352444841322442472867987730489.879.22420783113203491608040157180430360820541717335738404689526253152213919478250

66947905636248691862060483006 8.3574732949974249789435577946051867465589930465887918889188653634324782637654 6.880.36277589131552930

7313854445604079804026165243298890571460172146624950618253496832928674934214552747819636468.55798243333170729809441434699993769 35

22762753423163373320897881611581410703 5.881.5869837224475642233942560174083834122315824166562746183554346781470898130360035277192 3

85590837389538679371 5.213271538330595708752999229305180515911269321931108226250955151593636404835 81.882.949759613763093530708110463

012463216257747659955165189815606824772040431289500360647913493298291701687573 62.069069781663766438562440664005174285263545597273

42456358304493975479454590616.883.15367433362106577541023664804208466284893300766114398081710415029191302443255038883671520523820 3

06555555107.728234131999436214731543989331035480117481491920453278983400009134843095074197.884.2486502949973751284810476943433309 84

47470777365666049979866483276911706756150042490150655456803223572430843.93514111016581285858778805573155290864383605164779573534 17

04503110322549664813.885.4023246286184409038912843423854156473236407813178044806157689830610300919940508137382217598062353012798 59

51.6633752421652490733193320450625883887613175435682490143251045122451656447390 10.886.650974923615816032372332036728746631798348 55

4974464980414433815830147159555551238639728314374267537004167 95.598516352331061931907120100794141297405153595216044749666809015 35

5488194403823.887.1053299552234256936263616379114162279121989336292269461030202798891177251549602052377950074180502838284027 4.72618

91594496311005226452145856729686166471138784293763991913527600653839142833.888.1704274475850072968635948415842908910920337891266 73

44414446366147213244111051532910176783885547703752881954.286040794682737293713357224665087098357162473400033851365872254295614 2033

546656.889.27575740280843299048995647949570711900423272275590039024748394136125016626547553433956284627352732135722229.0122299541 32

36839423600243925076006697380958727846322776506360705567958726894 89.890.446184850393440287353551321079998010096266511882573834391 9

476028333826073759908634413306851290043588860418 3.2982707488151056879493596639158471653309720606784970791309358613512937906236145.

891.72194225320187327784350780057570512910049923463847422463943154419463277364146639778089353140253168024326412.310500702947474082

1853621031666072323047816479569603068959994684069733778925634.892.1168127103595313565197059121655703139196765746521048059031379147

0280153810174572612222242165315360391293059.56087714517625797701347217670824543976357537086354573860269353297582671685161779.893.1

8900693567971868430405669222314082682972649811595222836708106912226481546589236590031177479340677193725700.79192721547100538523200

838702490616299405353565924176929229347981652405464087413.894.30581964603925004082376260438871114074940307276805703427021898382506

635356763809202253419644656037585018760.35280436064726336224548056373315160275762890652278750789498701279235077149249192.895.49482

658171896872512781929661185196757912957088400926263730005294733116903353045792284597123996714778744461.144731576118268747477488950

75805776575168244218202927718728049260887482613336605.896.800646227758218765951581901000563108328532643652066296907519036772397522

60116854994538016768652752363763221.49753593676553210972296951449120936850931134870481678508226750540122559762585797.897.1295472809

4771874910794011976124150759076622145360755595448190897197286916346990078682261389264946714250768.26422675128838008572004584652492

6713426099379088684606226954799801010042375922402.898.2096119037235406257030983098612978184236194858188141856452338126492126214235

8675578136063066130221950627090.413980344964933296692342797974047650277030513959166284735181550341132602138508199.899.3391591846712

5937481103842962253932601438570727242174159971572162118549058705665656818324455395168664877858.678207096253313382412388644498974 36

3703129893047850890962136350142142644514430601.900.548771088394800000514136739483837144438005193091235927244949534270398112010643 4

1234954387521525390615504949.092187441218246679104731442473022013980160407007017175697317900483275246652938800.901.887930273066059

375325175169106376470452390900363657668844665255891583602597700068917727119769205592803828 07.770394537471560061517120086971996377 6

832903000548680666594542506254178911673694 01.902.143670136146085937583931190859021361489039609345489359608961479016198171460834348

12672709949844594989568877 5.686258197868980674062185152944501839166345070706188524235677215110869313782030820 1.903.23246316345269187

511644870776965900853427869938185512649342800460535653172060435501849981147536650917627056.463297651616136680213897161641701476934

67410071167533090162264017341110289876776 02.904.376133299598777812700379896628680370023318308727344486102389483621554703181438703 1

4522691097381245907215832.1495558494851173542760823145862033161010191714178638551372998552842804166807985803.905.608596463051469687

8168286063983393785575970081091996125958174882269112349020430581637267224491789682484288 8.612853501101254034489979476227904793035 6

93272129539186038922495457691519579566340 5.906.984729762650247500517208505027019748580915316836544098698206971848465938083481761 30

8953633422991427320587 20.762409350586371388766061790814108109136712443547403041176222350741971936260364920 8.907.1593326222570171718 8

3340371114253591271385123249457437112940244600753771729855248194726803558721703955690160.93752628516876254232560412670420129021724

0571567694222721514484619966345583993126l3.908.2578055988351964688851245616452378875719427641782287809992231431923843111069006580 7

816339892951618228896033.01376722022739968120221030578561210113091181592243452683913671969416353921002961821.909.41713822140536818 7

7185282727877738002857939966728031521286255891999220284054531400254314345167332218458619 3.951293505396162223527814432489813391348

152387490128749560651204314129884794022744 34.910.674943820240564656603652834433011687857736760851031933127848732392306339512353798

10359483344624940413482226.96506072562356190473002473827542549247906420341256327639978792400829342400405236255.911.109208204164593 2

844322181107220785488143530757523835085256474321592228367917806938129026267962982625980684 2.09163542310197241282578391707652388838

2721659090269202596043912832242330879807510689.912.176702586188649750092583394165379717600126751837486701838432305398453470743016 0

73623262110140923203011550 64.788141495664328603298786390904066437630628079431525530236022705233071673280212746944.913.2859107903532

4303452480150488745826641447982758987021036407973755767630753479676743616473693722146560961906.879776918766301016124570307980590 3

26013349738521794732832066618065314004160020257633.914.4626133765418927846173848990528379840146065794273569122025120429561297782 77

812841059426847078144668621169 71.667918414430629619423356698884656763643977817953320263068089323298385677440233004577.915.748524166

8951358191421864039402962504290864070172271225665917805138060858126096084955915840153661342307887 8.547695333196930635547927006865

2470896573275564751149959001559413636996816002532622 10.916.121113754343702860375957130299313423444369298644458403476910382346 99358

64090422449555018431093510802851 9585.021561374762756025497128370574990385330130537442843525896824526466208535904048 6266787.917.195

966171033216442290175770693343048487277939346181115733569560398374194990303205805061001510887693708274 72.876330908082449089051921 0

71261515094295863293090355025486840120602578504064073952 8997.918.317079925376919302666132900992656471931647237990639519210479942 74

536778139934545076056284462023877399347057.897892282845205114549049441836505479625993830533198551383664647068787039968122579578 4.9

19.513046096410135744956308671685999520418925177336820634944049503143741976389648656565623846131126467701 74530.7742231909276542036

00970513098020573921857123623553576870504767671365544032196532 4781.920.830126021787055047622441572678655992350572415327460154154 52

94458891097577889941073261866907513652416952158 8.672115473772859318150019954934526053547850954156752128254169414740152584000319 1' 2

0565.921.1343172118197190792578750244364655512769497592664280789098578949032851734178642763891810536882491709396961 1.944633866470 05

1352175099046803254662746970807778030570512467418241151812803251564453 46.922.217329813998424584020119181704331150512007000799174 09

432531083949219614919676368712179972276338569510921770.8118454' 38473372839901010422967072681017559031937057833378843597151670712 03

28347565911.923.3516470258181436632779942061407967017889567600656021732351687343954813226146279635109807764516348660489138 2.7564792

8031738863616520008909996193084872671097173635385035177795631888400653504011257.924.5689768398165682472981133878451278523009637608

6477626756047957388767747181139165063278049921502056115813153.5683246941647259201553011313966691989504826141654421371882361376 7148

595520981851577168.925.9206238656347119105761075939859245540899205209303784407956483082831587944260196141437612756666554272070 4536.

324803974482114556320501220496631129799209325137178491038587915627804839216353555884 25.926.1489600705451280157874220981831052406 39

0884281795154708356127882170836266237411264776541774881675988365176 8.98931286686468404764758023518933003287496919393026206282268 24

0532992907944261720716559 3.927.2410224571085992068450328575816976960480804802725533149151776190453995060663430878920303050548331 41

55722222.6217932643128955032796303572389931458548901264439799119265411968927095633642525627540 18.928.38998252765372722263245495576 4

8029366871689084520687857507904072624831326900842143696844825430007403937399 1.6111061311775795509272105924283231787298593203742419

7474922360222263864280686976991961 1.929.6310049847623264294774878133465006327352493887246221006659680263078826387564273022617147 87

5978338819509621 4.2328993954904750542068409496673163245847494468182218866757647991153482061711223326736 29.930.102098751241605365210

9942769111303569422418297176690886416758433570365771446511516631399270140834622344702 0.5844005526668054605134051542095639503314608

767192463861424988401337986848977992102593240.931.165199249717838008158743058245780420215766768590131298708272645987824841020293388

188931140577386685042956642.00769049221585296593408924917629558278993582140106857481007532004533350551491144352668 69.932.2672980009

594433733697373351569107771580085983078003873499484893448614181649450335524513327879503126640366 2.592091044882658426447494403385 8

5953312139669812031496095257416017913219041271065378601 09.933.432497250677281381528480393402691197373775366897931686058221135332 68

62591852389154417627385618171630936030 4.5997815370985113923815836525621551159113325195213835357626494802244656959276220973126978.9

34.699795251636724754898217728559601974531783965205732073408169624677547677350183948994214071349767475757639 67.191872581981169 8188

29078055948014649032729217641698496715223640403597886340332751098708 7.935.113229250231400613642669812196229317190555933210366375 94

66390760010233936535422864435976809911584638851242 7.17916541190796812112106617085101697649440617371630820324778731206280635822679 5

48484114065.936.18320877539507308913249158505218951464373432973093958328745603846877816138856068134301908812613521146088823.898352 6

70106085103003973976445818441397679095480478052919309676103166146860828759951011 52.937.29643802562647370277516139724841883183429 02

629413059592340951144698015550421029677866167691172936753460125 1.07751808201405322412504014729683541789208526919678625616709698816

59725050876242447921521 7.938.47964680102154679190765298230060834647802459267224554252155115293857971643066364912963585724342888680

690074.97587075212013832712901412374265385928976436467726430908640666426913865194845300474316369.939.77608482664802049468281437954

902717831231485561355150175564626740838127147276661691625262636072256215291326.05338883413419155125405427103948927718184963387405

056525350365243511115703607724953531586.940.12557316276695672865904673618496355247903394482857970442771974203469609879034302660458

8848360415144895981140.102925958625432987838306839478214313647161399855131487433991031670424980898453025427647955.941.2031816454317

5877812732817413986627031026543038993485460328436877553422593761968829621411099648740111127272.70826484203885214296371226658216324

1365346363242536543959341396913936096602060750381379541.942.3287548081987155067863749103248298227892993752185145590310041108102303

2472796271490080295935690254600725412.811190800664285130802019106060377555012507763097668031393332428584361077500513775809227496.9

43.53193645363047428491370308446469609309956480560844941363428847958576455066558240319701707035338994711852685.5194556427031372737

65731372642540796377854126340204575352673825498297174102574526190607037.944.860691261829189791700077994789525915888864180826963972

66529259039599487539354511809782002971029249312578098.3306464433674224045677504787029183513903618894378726067460062540826582516030

88301999834533.945.13926277154596640766137810792542220089884289864354133862995810699817594260591275212948371000636824402443078.385

0102086070559678333481851345459147768216015778077182098680079580955425705662828190441570.946.2253318977288853868313859074043747924

8772931672623773589648736603777543014526726393926571297739749333700888.21807485294379820829012323300483774991585779052159497888844

686333663613677308751130190276103.947.36459466927485179449276401532979699338657221536977907452644547303595137275118001606874942298

376573736143966.6030850615508541761234714181393836646926793920994026970943366413244569103014413958380717673.948.58992656700373718 1

3241499227341717858743015320960168104229328390737268028964472800080151359611632306984485 4.821159914494652384413594651144221414 6085

3718262099767597880527469081827803231650885709937 76.949.95452123627858897581691393806396877926087374746579588494937831210967817564

76272960767645589449289680598882 1.4242449760455065605370660692836050793012165747204003730731419160152751883337579046951711449.950.1

5444478032823261571410638607981405651351752795618126953723111511834049785440745760847796949060921987583367.624540489054015894495 06

607204278264939097537573413980490519471907060934663660744135522705225.951.2498969039560915132957977798862109344396049027027608580 3

2168946329308315419170187216154425385102116681822 49.766964986658566550548772678971143157321097033206179842212508910672136865469983

23182474416674.952.40434168428432412900990416596602499095312243065894212756940006144764881327357764482463239487571133655765617.3915

054757125824450438387510139258067120724089403196471177036297427462121065906731799712189 9.953.65423858824041564230570194585223592 53

9272733361702985601569007776957128692747832040786820260813453239478 67.158470462371148995592611429985068964033169442146499489330 21

254041488307757657390500471538573.954.10585802725247397713156061118182609163458497640206451131709690692246059419663254768654192151

365247897971348.454997593808373144063645018099899477074524185108681913644791617015762928968316457818468660472.955.1712818860765155

4136213080576704968417385770973823480987725380770015630706590733089062060353973382430366135.1708446400454880436229061610984063734

77841129323331862577812871057251236725973848318940199045.956.277139913328989518493691416948875775808442686140299321194350714622616

90126253987857716252505338630328337483.6258422338538611876865511791983058505523653144320137762226044880730141656942903061374088595

17.957.44842179940550505985582222271592545998230039587853413107160452232277320832844720946778312859312012758703618.796686873899349

231309457340296712224030206443755345638800417359130265402420264154456349058562.958.72556171273449457834951363966480123579074308201

883345226595523694539010959098708804494565364650643087041102.422529107753210418996008519495018074582571758187359415023021847203279

568114554460593757918079.959.11739835121399996382053358623807266957730434778973675833375597592681633179194342975127287822396265584

574472.1219215981652559650305465859791730298612778201942705053823439206333544970534818615050106976641.960.189954522487449421655484

95020455279315637865599162010356035149962135534275104213855576744358861329893278582.364174508940577006930147437928674837319534996

0130064468846461053536824538649373075643864894720.961.3073528737014493854760185364426254627336830037813568618941074755481716745429

85568307040321812575954778530544.860961071058329719606940239078478671808128162072769522669900259870369509184191690693971871361.962

.497307396188898807131503486647178255890061659772976965454458975169527017294027706862807765401189253711316 36.85027061604640997889 0

8414618365227045003478122202833991516361313407194047833564766337836766081.963.804660269890348192607522023089803718623744663554333 8

27348566450717698691837013275169848087213765208489846 91.336366723152242950851535485744370571681160628427560351418626157327756355 70

17756457031808637442.964.130196766607924699973902550973698197451380632332731079280302542588722570913104098203265585261495446220116

32.8186637339198652929742376947580893276181508440647843750570262288668475760485132122336964540352 3.965.21066279359695951923465475 3

28267856931375509868816446201515918766049244009680542572025039398287196706910101.952300406235089588059391243332526384786266906907

540410198888844599623211618690776804014540409 65.966.34085956020488421920855730425637676676513573102089554129546173024921501100990 9

523923515979244367413289217 34.77096414015495488103362893809061571240441775097232478525591507346647078766720398903771099444488.967

55152235380184373844321205753905533607889082970906000331062091790970745110671494964376637322723938035831836.72326454639004446909 30

2018142314209719068465787986519545480391806609399928589476584172553485453.968.892381914006727957651769361795432102844026560729955 5

446060826481589224621166244735672823524716067936475357 1.494228686544999350126649119513757809595102408852189980710718991532564786 95

30987548794365292994 1.969.1443904267808571696094981419334487438922917390439015547916703566068629913223339423211048725698846 1740058

540.821749323293504381921966930093689990678578706673205517616552290959865878623899352072116206415394.970.233628618181529965374 67507

811299195417669439511689710925227862142275523753399638967783310781704529676533897.97117219194800431693463184204506577 1638088947558

424515687624190113122357319209227560059859345335.971.378019044962387134984173220046440698068986134160798664043948978029618228 85633

033199893798038692991416592438.79292151524150869885659877213875576231666765423163003330417648107298823594310857963217605576 0729.97

2.611647663143917100358848298159432652245680529277695773296227599452373466390326721676771088203975210931263 36.764093707189513 01579

123061418382153395475660179005454899180067118611059326231780719223592510 6064.973.989666708106304235343021518205873350314666663 4384

944373401765774819916952466570536757090685909051250971877 5.557015222431021714647829386322577296271424256021684582295977152 2590988

292054263868244119908667 93.974.160131437125022133570186981636530600256034719271619021063640417693436516163698377535248015679488033

60284511.232110892962053473043906000050639883022618085781173913128777782344520942246774419401664791597285 7.975.259098107935652557 10

448913345711793528750138561546846479765807544163568568836408290281892253857854611256388.787812415205155644508688938682897612649760

511383342371358375497570430825167317058084105990683965 0.976.41922954506067469067467611509364853554353610488708748586129849313 50722

01852062460438066938218066579715409 00.0199233081672091175525949387335374956723785971645162844871532799149517674140914774857707822 8

12507.977.67832765299632724777916524855076647083103749050255595065895656857670788754042654334088586075664512582797288.807735723372

364762061283877416435108322139108547858655845528777485382592581408535569876772965215 7.978.1097557198057001938453841363644415006374

57359538964343652025506171117800893924890037789527989747111705543381 8.88276590315395738796138788161499726039945177057123749403326820

574003343599955000130556475552464664.979.177588485105332918623300661219518147720561108589219938717921163028848797693291554711983 86

597313568313713547.763539475491193864167516269356640771231685681426023359617821083488571695257690854862552432821168 21.980.287344204

91103311246868479758395964835801846812818428236994666920002680663254045508987914587060685369147366.64630537864515125212890415097 16

3803163111745199726085365108928922860513125724085616811718834581485.981.46493269001636603109198545880347779607857957671740422108 78

678322288756043258320098018630118437425368286091 4.4098448541363451162964204203282788028627831334232842132689103727171768265149317 1

1030669621166983 06.982.752276894927399143560670256387437444436598044845588503457814501428902410958372464891742157714349390520082 81.

0561502327814963684253245712999168344939005854205450669199996619457819577721725671987868095127979 1.983.121720958494376517465265571

519091524051517762156299272454568233365777801528420447469360516955809192734869 19.546599508691784148472174499162819563735668371884

382928018891003466295878428710427822945643067978097.984.19694864798711643182133259715783526849517756664085812280034968350866804262

425769395853473272724413178687747.652214531969933785314706956292811247185058430426437434710890969660874074205927684542824324019257

888.985.31866960648149294928659816867692679254669532879715739525491791687444584415267814142789524968305332452174667.19881404066171

7933786881455455630810920726802310820362729781973127169952634638112365769967087235985.986.5156182544686093811079307658347620610418

7289543801551805526760038311388677693583538642998241029745630862414.85102857263165171910158841174844205810578523273725779744067294

2788044026840565796908594291106493873.987.8342878609501023303945289345116888535885682242351729133101855172575597309296139768143252

3209335078083037082.049842613293369652888469867204072869026512035048078160170454915915213979475203909274364258193729858.988.134990

61154187117115024597003464509146304411196731884313654531176406736177065498122007552145036482371389949.6900871185925021371990058278

952514927132297267785335957611127858703258006315769706182958549300223731.989.21841939763688140418969886348581397682190093439083613

446756386348982333486361637890150804465969990179693657.895071379921839102487852814615658779615880930283341411778158277461847198579

0973615457322807493953589.990.3534100091787525753399448335204590682849450463581549776041091752538906966342713601215835661100647255

1083607.5851584985143412396868586425109102723291106570618750075392710633321729992106743321640281356794177320.991.571829406815633979

5296436970062730451068459807489911120716730387437140314978877390230916107697646273077726 5.48022987843618034217471145712656905194 49

9158734521641931742934079402019778977169370976041642881309 09.992.92523941599438655486958853052673211339179102710714608967578221399

760472813215909914467517687982935281860873.065388376950521581861570099637479324274102244407091426856700404126193197000446025873 78

85521082308229.993.14970688228100205343992322275330051584986370078561372017474552527413187596300468381677667876495939801263813.854

561825538670192403628155676404837621909383175230784617412974492021339479021771958354896853704391 38.994.24223082388044070892688 2075

8059737271890428034963283291423237466738923487762205937312441964529423332944990 1.161100663233722350589785165640152770049319607615 9

39927303083014904640659179066374545733752064527473 67.995.39193770616144276236680529855927424303890650428194204931706927194802422 47

392252775480208752179017313071371 5.015662488772392542993413321316557607671228990791170711920495989396661998658088146504088648918 23

186505.996.6341685300418834712936873743652479702279493077782703784593930186219165735154458712792650716708440646016361 6.17676315200

61148935831984869567103777205485984071106392235790043013026578371545210498224009827593387 2.997.10261062362033262336604926729245222

132668558120602124277764622905699407982546711488272859468887457959087733 11924256407785074365766118082732679853917775891982813511 4

4074993697964656495242667553911049900991203 77.998.16602747662452097049541800472897701834948051198384828062358553091918573717701170

201065510185595898605104094.736918879278462233015981029522997836311232618760539199036765399799926731433239718860373345088375054249

.999.26863810024485359386146727202142923967616609318986952340123175997617981700247881689338369654483356564191827.85616144335631297

6673642210350324634850410377680367334151172899169723197082763985615764450078474174626.1000.434665576869374564356885276750406258025

6466051737178040248172908953655541794905189040387984007925516929592.25930803226347752096896232398733224711616429964409065331879382

98969649928516003704476137795166849228875.1001.70330367711422815821835254877183549770181269836358732742604905087154537118196933579

74224949456261173348775.0449241765991088186363265450223647106012053374121273867339111198139373125598767690091902245245323403501.

Made in the USA
Monee, IL
04 November 2026

67377504R00046